CHASING EXTREME WEATHER

CHRISTINE HONDERS

PowerKiDS press.

NEW YORK

Published in 2019 by The Rosen Publishing Group, Inc.
29 East 21st Street, New York, NY 10010

Editor: Elizabeth Krajnik
Cover Design: Michael Flynn
Interior Layout: Reann Nye

Photo Credits: Cover (background) Cammie Czuchnicki/Shutterstock.com; cover (truck), p. 10 Drew Angerer/Getty Images News/Getty Images; pp. 3–24 (background) chaowat kawera/Shutterstock.com; p. 4 John Normile/Getty Images News/Getty Images; p. 5 Ryan McGinnis/Moment/Getty Images; p. 7 JOSH EDELSON/AFP/Getty Images; pp. 8, 11 Barcroft Media/Getty Images; pp. 9, 14 Courtesy of NOAA's Photo Library; p. 13 Willoughby Owen/ Moment/Getty Images; p. 15 Eddie J. Rodriquez/Shutterstock.com; p. 17 https://www.nasa.gov/sites/default/files/thumbnails/image/image1-irma-goes-sunday.jpg; p. 18 bibiphoto/Shutterstock.com; p. 19 Scharfsinn/Shutterstock.com; p. 21 Newspix/Getty Images.

Library of Congress Cataloging-in-Publication Data

Names: Honders, Christine, author.
Title: Chasing extreme weather / Christine Honders.
Description: New York : PowerKids Press, [2019] | Series: Spotlight on weather and natural disasters | Includes index.
Identifiers: LCCN 2017060307| ISBN 9781508168768 (library bound) | ISBN 9781508168782 (pbk.) | ISBN 9781508168799 (6 pack)
Subjects: LCSH: Storm chasers--Juvenile literature. | Severe storms-- Juvenile literature. | Natural disasters--Juvenile literature.
Classification: LCC QC941.3 .H66 2018 | DDC 551.55--dc23
LC record available at https://lccn.loc.gov/2017060307

Manufactured in the United States of America

CPSIA Compliance Information: Batch #CS18PK For further information contact Rosen Publishing, New York, New York at 1-800-237-9932.

CONTENTS

EXTREME-WEATHER EXPERTS 4
CHASING DANGER 6
TORNADOES 8
FINDING THE PERFECT STORM 10
STORM-CHASING GEAR 12
EXTREME-WEATHER RESEARCH 14
HURRICANE HUNTERS 16
SNOW SCIENTIST 18
EXTREME WEATHER ON THE RISE 20
WHO CHASES EXTREME WEATHER? 22
GLOSSARY 23
INDEX 24
PRIMARY SOURCE LIST 24
WEBSITES 24

CHAPTER ONE

EXTREME-WEATHER EXPERTS

Everyone on Earth experiences weather. In some places, the weather is hot and sunny for most of the year. In other places, it rains for months at a time. We get used to dealing with the kind of weather that happens where we live. People in warm climates may experience many thunderstorms, while people in cold climates may experience many blizzards. Sometimes the weather can be scary and even dangerous.

Meteorologists at the Center for Severe Weather Research (CSWR) study tornadoes and hurricanes. The CSWR has vehicles that help meteorologists collect data about severe weather.

Experts use the words "extreme weather" when talking about weather that's dangerous to people or property. These experts study the weather and try to **predict** when and where extreme weather may strike. While most of us try to get away from dangerous weather, extreme-weather experts try to get as close to it as they can. They chase extreme weather to learn more about why it happens and how we can prepare for it.

CHAPTER TWO

CHASING DANGER

Why would someone put himself or herself in harm's way to chase extreme weather? Most extreme-weather chasers are meteorologists or other scientists who gather data about extreme weather so they can understand and predict it. Some extreme-weather chasers are photographers who sell pictures of extreme weather to make money. Other people like the thrill and excitement that come with seeing extreme weather events up close. However, unless you're an expert who's studied extreme weather, this isn't a good idea. Many people have died while chasing extreme weather.

Real extreme-weather chasing isn't always as exciting as it looks on TV. These people often spend many hours studying extreme weather, planning where to find it, and making sure their equipment is working. Even after all this planning, sometimes chasers drive around for hours and hours and never see a major storm.

Extreme-weather chasers use video cameras, digital cameras, laptops, **radiosondes**, and chaser vehicles to figure out where extreme weather will strike.

CHAPTER THREE

TORNADOES

Tornadoes are **violent**, spinning funnels of air. They often form from an uncommon type of thunderstorm called a supercell. Tornadoes span from the clouds to the ground and have wind speeds that reach up to 300 miles (482.8 km) per hour. They can destroy buildings, uproot trees, and throw cars hundreds of feet through the air, causing millions of dollars in damage and even loss of life. There are about 1,200 tornadoes reported in the United States every year, many of which occur in the central part of the United States, in an area known as Tornado Alley.

SUPERCELL THUNDERSTORM

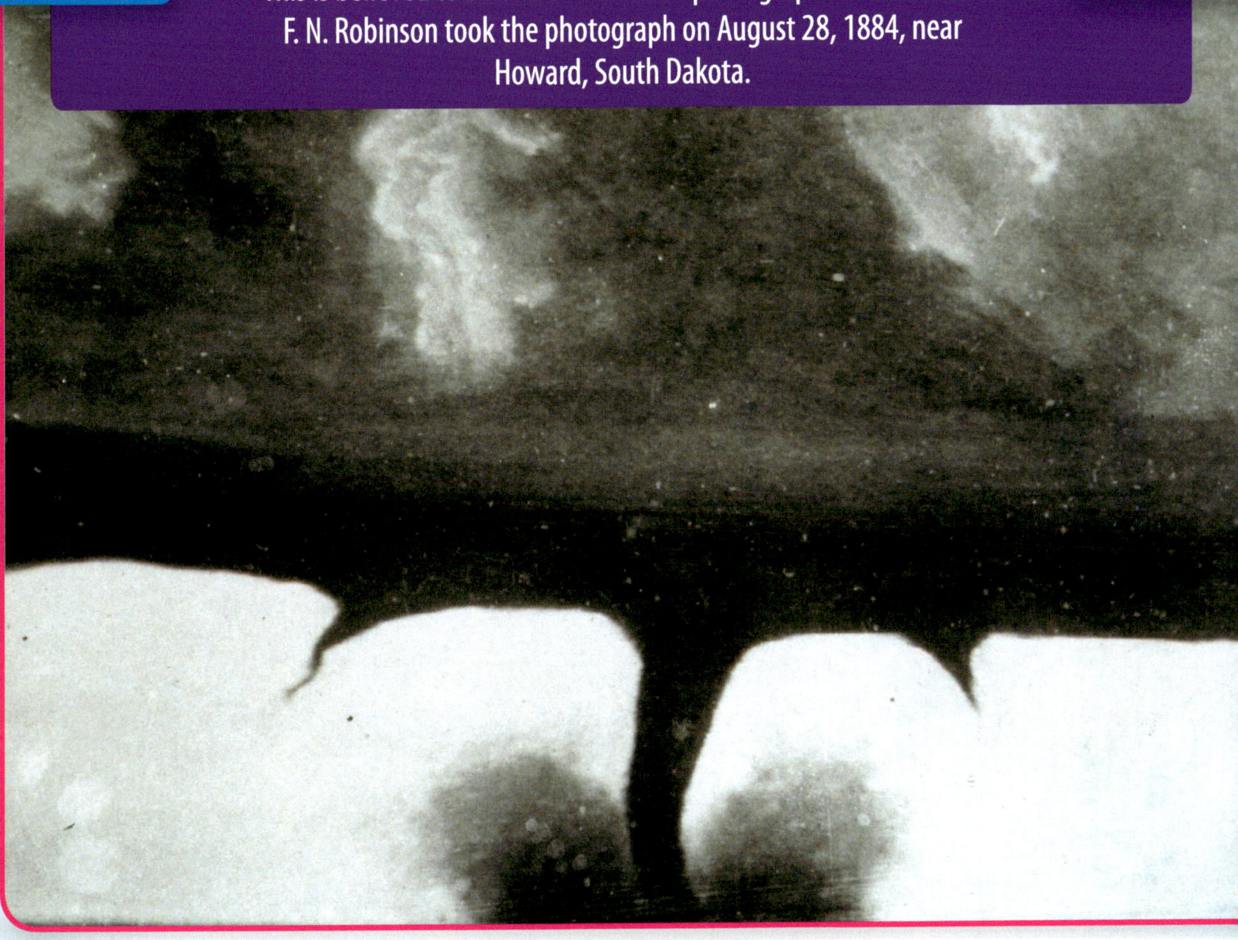

This is believed to be one of the oldest photographs of a tornado. F. N. Robinson took the photograph on August 28, 1884, near Howard, South Dakota.

In the United States, a tornado can form when warm, moist air from the Gulf of Mexico meets cold, dry air from Canada. When they meet, a change in wind direction and an increase in wind speed can cause the air to spin in a horizontal column. As the air rises, it tilts the spinning column upward, creating a tornado.

CHAPTER FOUR

FINDING THE PERFECT STORM

Tornado-chasing season in Tornado Alley runs from spring to summer. The night before and morning of a chase, chasers check reports from the National Weather Service (NWS). They look for cold air at high **altitudes**, warm air near the ground, and wind blowing in different directions. When they find a location where they think a tornado could form, they start driving toward it.

Byron Turk, a photographer who worked on the TV show *Storm Chasers*, says tracking tornadoes requires people to make many correct decisions to find a worthwhile tornado.

On the way, chasers use computers, phones, and radios to receive up-to-date weather information. They talk to SkyWarn spotters, who are trained to report on storms. If a storm suddenly changes direction, chasers change direction too. Chasers look for signs of tornadoes, such as clouds that spin counterclockwise, which may be a sign of a supercell. Then, they find a place where they can safely set up their equipment and still get a great view.

CHAPTER FIVE

STORM-CHASING GEAR

Storm chasers use basic gear such as video cameras, laptops, and radios. However, to locate tornadoes and find out what happens inside them, they also use special **technology** to track storms.

CSWR's Doppler on Wheels (DOW) network is a group of three mobile **radar** trucks, four support vehicles, a mobile office and repair van, and weather stations that are carried in the support vehicles. Doppler radar is a system used especially for predicting the weather. Storm chasers use this gear to get as close to a weather system as possible and take measurements used to create detailed weather images.

Tornado **Intercept** Vehicles (TIVs) are heavy, armored trucks built to withstand direct contact with a tornado. TIVs are built with special supplies to prevent damage from a tornado. Onboard each existing TIV is an IMAX camera that can take 360-degree footage, allowing filmmakers to film what happens near (and even inside) a tornado from all directions.

The TIV1 started out as a pickup truck. Today, it has a body and floors made of steel plates and strong plastic windows to prevent damage from flying objects. The TIV2 (shown here) is similar to the TIV1, but has slightly different features to make it an improvement on the TIV1.

CHAPTER SIX

EXTREME-WEATHER RESEARCH

While we've learned a lot about tornadoes, there's still a lot that remains unknown. Tornadoes only occur under the right conditions, they can change course quickly, and they often cause many deaths. The **research** done at the U.S. National Severe Storms Laboratory (NSSL) helps meteorologists improve tornado **forecasts**, which help warn people and save lives.

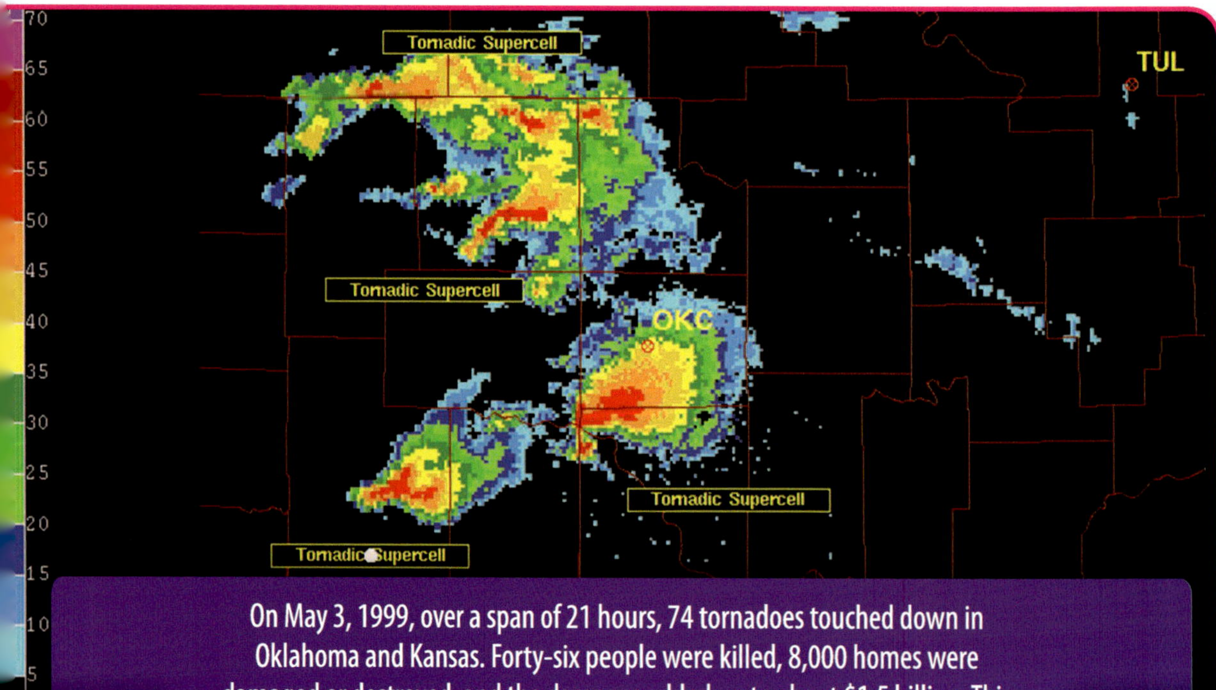

On May 3, 1999, over a span of 21 hours, 74 tornadoes touched down in Oklahoma and Kansas. Forty-six people were killed, 8,000 homes were damaged or destroyed, and the damages added up to about $1.5 billion. This radar image of central Oklahoma shows four supercells at the same time.

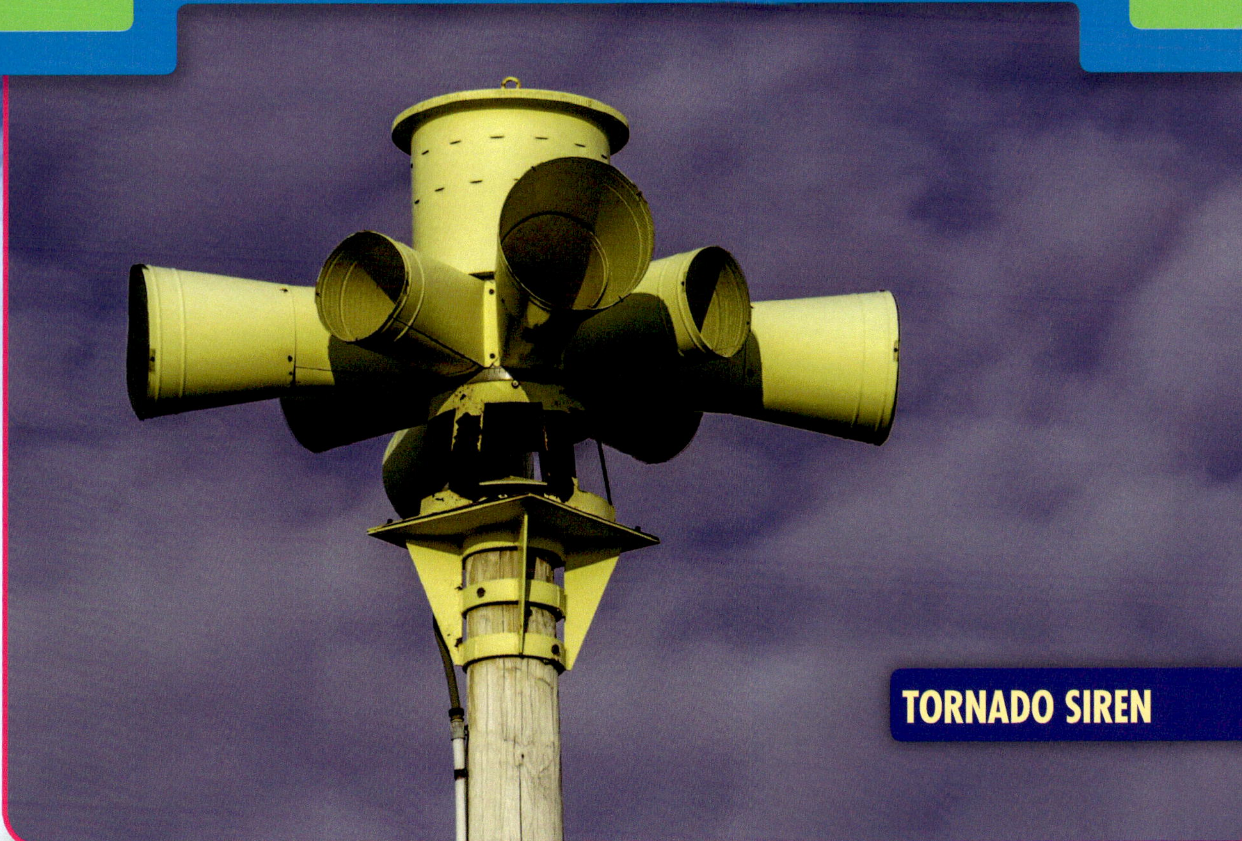

The NSSL has created a computer model that uses 3-D imaging to **simulate** thunderstorms that cause tornadoes. This helps researchers understand what conditions must be present to cause a tornado to form. Researchers at the NSSL also use a special type of radar that searches the sky for extreme weather five times faster than other radars. NSSL's Warn-on-Forecast project aims to use computers to predict where tornadoes will form within an hour so people have more time to prepare.

CHAPTER SEVEN

HURRICANE HUNTERS

Hurricanes are large rotating storms that form over warm tropical oceans. They form when warm, moist air rises, cools, and becomes clouds. The cooler air that replaces it also warms, rises, and joins the clouds, creating huge storm clouds that rotate. As the hurricane moves over the warm ocean, the winds get stronger, sometimes reaching speeds as high as 155 miles (249.5 km) per hour or more. Such strong winds can destroy homes and kill many people.

The National Oceanic and Atmospheric **Administration** (NOAA) has a group of hurricane hunters, which are special aircraft that travel into hurricanes. Scientists release **dropsondes** that record the temperature, wind speed, and humidity, or the amount of moisture in the air, from inside the storm. Other aircraft fly high in the sky to show scientists what hurricane development looks like from a higher altitude. This data helps NOAA predict how strong a hurricane might be and when it will hit land, which can save lives.

Hurricane Irma is the strongest Atlantic hurricane on record. It damaged parts of the Caribbean Islands, Puerto Rico, and southern Florida, killed dozens, and left millions without power.

CHAPTER EIGHT

SNOW SCIENTIST

Winter weather in many regions can be dangerous due to very low temperatures and low visibility. It can leave people without water or power for days at a time. There are different types of winter storms, including blizzards, ice storms, and lake-effect storms. Winter storms are a combination of heavy snow, strong winds that blow snow around, and temperatures so low that it may be unsafe to be outside.

Blizzards have winds of at least 35 miles (56.3 km) per hour with falling or blowing snow. Visibility must be less than a quarter mile (.4 km) for at least three hours for a storm to be considered a blizzard.

The U.S. Department of Agriculture employs people to collect and study data on water and snowfall at different sites throughout the western United States and Alaska. These people measure snow depth, temperature, and soil moisture to forecast water supply. This information is very useful when forecasting extreme weather because it shows how the data for one year follows the data from previous years.

CHAPTER NINE

EXTREME WEATHER ON THE RISE

Earth's climate is changing. In recent years, cases of extreme weather have started happening more often. Heat waves are longer. Some locations experience larger amounts of precipitation, causing flooding. At the same time, other areas have longer, more severe droughts, which are long periods of little to no precipitation. While scientists say some of Earth's weather patterns may be changing naturally, human activities are the main cause of climate change.

In the Climate Science Special Report published in 2017, scientists at 13 federal agencies explain how climate change has caused extreme weather events, especially extreme temperatures, and that humans are a large part of the problem. In this report, researchers examined a record-breaking heat wave in Australia in 2013 and determined that this type of extreme weather wouldn't have occurred if not for human-caused climate change.

Australia experienced its hottest spring and summer on record in 2013. Wildfires can be one of the biggest issues during a heat wave. Fires may burn uncontrollably for days because the earth is so dry.

CHAPTER TEN

WHO CHASES EXTREME WEATHER?

Real extreme-weather chasers are scientists. They study meteorology and research weather patterns and climate. Many of them study earth science and **engineering** in college. Dr. Joshua Wurman designed the Doppler on Wheels system. Tim Samaras, a scientist who became famous as part of the TV show *Storm Chasers*, spent 35 years keeping track of tornadoes and died while chasing a tornado. Tim Marshall saw a tornado when he was 10 and wanted to be a meteorologist and an engineer when he grew up. Now he is. His job is to go to places after they've been hit by extreme storms to look at the damage and help people rebuild.

People who chase extreme weather put their lives in harm's way in the name of science. They hunt storms because they want to know why they happen and how they can predict them so the rest of us can stay safe.

GLOSSARY

administration (ed-mih-nuh-STRAY-shuhn) A government board or organization.

altitude (AL-tuh-tood) The height of something above a certain level.

dropsonde (DRAHP-sahnd) A radiosonde dropped out of an aircraft to collect data about the air.

engineering (en-juh-NEER-ing) The study and practice of using math and science to do useful things, such as building machines.

forecast (FOR-kast) A prediction of future conditions. Also, to predict future conditions based on data.

intercept (in-tuhr-SEPT) To take, reach, seize, or stop something before it reaches an intended destination.

meteorologist (mee-tee-uh-RAH-luh-jist) Someone who studies meteorology, or the science that deals with the atmosphere, weather, and weather forecasting.

predict (prih-DIHKT) To guess what will happen in the future based on facts or knowledge.

radar (RAY-dahr) A machine that uses radio waves to locate and identify objects.

radiosonde (RAY-dee-oh-sahnd) A very small radio transmitter carried into the sky to collect weather data.

research (REE-surch) Careful study that is done to find and report new knowledge about something.

simulate (SIM-yuh-layt) To represent the operation of a process by means of another system.

technology (tek-NAH-luh-jee) A method that uses science to solve problems and the tools used to solve those problems.

violent (VY-uh-luhnt) Showing very strong force.

INDEX

A
Alaska, 19
Australia, 20, 21

B
blizzards, 4, 18, 19

C
Canada, 9
Caribbean Islands, 17
Center for Severe Weather Research (CSWR), 5, 12

D
Department of Agriculture, U.S., 19
Doppler on Wheels (DOW), 12, 22

F
Florida, 17

G
Gulf of Mexico, 9

H
hurricanes, 5, 16, 17

K
Kansas, 14

M
meteorologists, 5, 6, 14, 22

N
National Oceanic and Atmospheric Administration (NOAA), 16
National Severe Storms Laboratory, U.S. (NSSL), 14, 15
National Weather Service (NWS), 10

O
Oklahoma, 14

P
Puerto Rico, 17

R
radar, 12, 14, 15

S
scientists, 6, 22
SkyWarn, 11
snow, 18, 19
South Dakota, 9
Storm Chasers, 11, 22
supercell, 8, 11, 14

T
thunderstorms, 4, 8, 15
Tornado Alley, 8, 10
tornadoes, 5, 8, 9, 10, 11, 12, 14, 15, 22
Tornado Intercept Vehicles (TIVs), 12, 13
Turk, Byron, 11

U
United States, 8, 9, 19

V
vehicles, 5, 7, 12

W
Warn-on-Forecast, 15
wind, 8, 9, 10, 16, 18, 19

PRIMARY SOURCE LIST

Page 9
Tornado 22 miles southwest of Howard, South Dakota. Photograph. Taken by F. N. Robinson. August 28, 1884. NOAA's National Weather Service (NWS) Collection. http://www.photolib.noaa.gov/htmls/wea00206.htm

Page 14
Radar image of central Oklahoma. National Weather Service. May 3, 1999. https://www.nssl.noaa.gov/about/history/may3rd/images/supercells.gif

Page 17
Satellite photo of Hurricane Irma. Photograph. Taken by the NOAA GOES East satellite. September 10, 2017.

WEBSITES

Due to the changing nature of Internet links, PowerKids Press has developed an online list of websites related to the subject of this book. This site is updated regularly. Please use this link to access the list: www.powerkidslinks.com/swnd/extreme